WALCH  PUBLISHING

Daily Warm-Ups

SAT PREP
MATH

Maureen Steddin

Level II

The classroom teacher may reproduce materials in this book for classroom use only.
The reproduction of any part for an entire school or school system is strictly prohibited.
No part of this publication may be transmitted, stored, or recorded in any form
without written permission from the publisher.

1   2   3   4   5   6   7   8   9   10

ISBN 0-8251-5891-5

Copyright © 2006

J. Weston Walch, Publisher

P.O. Box 658 • Portland, Maine 04104-0658

walch.com

Printed in the United States of America

Table of Contents

Daily Warm-Ups: SAT Prep—Math

The *Daily Warm-Ups* series is a wonderful way to turn extra classroom minutes into valuable learning time. The 180 quick activities—one for each day of the school year—practice the test-prep skills necessary for success on the math portion of the SAT. These daily activities may be used at the very beginning of class to get students into learning mode, near the end of class to make good educational use of that transitional time, in the middle of class to shift gears between lessons—or whenever else you have minutes that now go unused.

Daily Warm-Ups are easy-to-use reproducibles—simply photocopy the day's activity and distribute it. Or make a transparency of the activity and project it on the board. You may want to use the activities for extra-credit points or as a check on the math and test-taking skills that are built and acquired over time.

However you choose to use them, *Daily Warm-Ups* are a convenient and useful supplement to your regular lesson plans. Make every minute of your class time count!

Reference Information

Area Facts

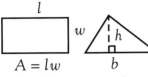

$$A = lw$$

$$A = \frac{1}{2}bh$$

$$A = \pi r^2$$
$$C = 2\pi r$$

Volume Facts

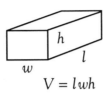

$$V = lwh$$

$$V = \pi r^2 h$$

Triangle Facts

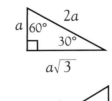

$$a^2 + b^2 = c^2$$

Angle Facts

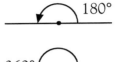

180°

360°

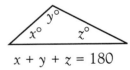

$$x + y + z = 180$$

Response Grid

Some questions include a response grid rather than a multiple-choice answer grid. Students should write the answer in the blank spaces at the top of the grid. Then they should fill in the circles that correspond to the digit or symbol. Some examples are shown below.

Answer = 100
Both placements are acceptable.

Fractions or decimals are shown as above.
Change mixed numbers to fractions or decimals.

Daily Warm-Ups: SAT Prep—Math

Daily Warm-Ups: SAT Prep—Math

Numbers and Operations

Martha is dividing cookies into packages. If she puts 5 cookies in each package, she will make 33 packages with no cookies left over. If instead she puts 3 cookies in each package, how many packages can she make?

(A) 11

(B) 41

(C) 55

(D) 99

(E) 165

1

    (A) (B) (C) (D) (E)

Numbers and Operations

The weight of 80 screws in a jar of identical screws is 2 ounces. What is the weight, in ounces, of 4 screws?

(A) 0.05

(B) 0.1

(C) 0.5

(D) 1

(E) 10

2

Ⓐ Ⓑ Ⓒ Ⓓ Ⓔ

Numbers and Operations

All numbers that are divisible by both 3 and 5 are also divisible by 10.

Which of the following numbers can be used to show that the statement above is FALSE?

(A) 20

(B) 25

(C) 30

(D) 40

(E) 45

Numbers and Operations

Gabo bought three chicken dinners for $6.95 each and one hamburger meal for $5.75. He and three friends decide to divide the cost of the meals equally. How much money does Gabo receive in total from his three friends?

(A) $6.35

(B) $6.65

(C) $19.65

(D) $19.95

(E) $26.60

4

Numbers and Operations

What is the least positive integer divisible by the numbers 2, 4, and 7?

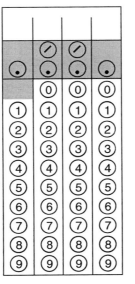

The blocks in a certain toy chest are either red or blue. If the ratio of the number of red blocks to the number of blue blocks is $\frac{1}{3}$, each of the following could be the number of blocks in the chest EXCEPT

(A) 12

(B) 16

(C) 20

(D) 30

(E) 36

6

© 2006 Walch Publishing

Numbers and Operations

If it is now Thursday, what day will it be 200 days from now?

(A) Sunday

(B) Monday

(C) Wednesday

(D) Thursday

(E) Saturday

A store sells its merchandise at 20% above the wholesale price. If the store sells an item for $30.00, what is its wholesale price?

(A) $6.00

(B) $24.00

(C) $25.00

(D) $27.00

(E) $36.00

8

   Ⓐ Ⓑ Ⓒ Ⓓ Ⓔ

Numbers and Operations

Greg bought 6 bottles of water from a vendor. He paid with a ten-dollar bill and got back two dollars in change. He realized that he had received too much change and gave one dollar back to the vendor. What was the price of each bottle of water?

(A) $1.00

(B) $1.15

(C) $1.25

(D) $1.50

(E) $2.00

$$\frac{\square - 3}{4} = 2\frac{1}{2}$$

What number, when used in place of $\square$ above, makes the statement true?

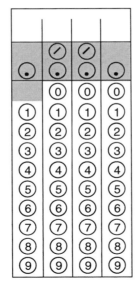

Daily Warm-Ups: SAT Prep—Math

Numbers and Operations

If Maia can read 20 pages in 18 minutes, how long will it take her to read 150 pages?

(A) 1 hour and 6 minutes

(B) 1 hour and 35 minutes

(C) 2 hours and 15 minutes

(D) 2 hours and 46 minutes

(E) 3 hours and 18 minutes

    (A) (B) (C) (D) (E)

11

At North High School, 66 students are in the band and 45 students are in the orchestra. If 24 students are in both band and orchestra, what is the ratio of the number of students who are only in the band to the number who are only in the orchestra?

(A) 2:1

(B) 21:12

(C) 18:11

(D) 8:7

(E) 14:15

12

Ⓐ Ⓑ Ⓒ Ⓓ Ⓔ

Numbers and Operations

$X = \{3, 6, 9\}$ and $Y = \{4, 8, 12\}$. If a is in X and b is in Y, how many different values are there for $\dfrac{a}{b}$?

(A) 3

(B) 6

(C) 7

(D) 8

(E) 9

Ⓐ Ⓑ Ⓒ Ⓓ Ⓔ

13

© 2006 Walch Publishing

Numbers and Operations

All of the following can be expressed as the sum of two consecutive integers EXCEPT

(A) 27

(B) 30

(C) 33

(D) 39

(E) 43

Ⓐ　Ⓑ　Ⓒ　Ⓓ　Ⓔ

Numbers and Operations

The number n is a multiple of 3 and is a factor of 48. Grid a possible value for x.

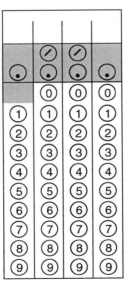

Numbers and Operations

Ginny is writing a report for her science class. Her teacher says that all numbers in the report must be written in scientific notation. If Ginny wants to include the number 5,067,000,000, how must she write it in her report?

(A) 50.67×10^6

(B) 50.67×10^8

(C) 5.067×10^8

(D) 5.067×10^9

(E) 5.067×10^{10}

16

   (A) (B) (C) (D) (E)

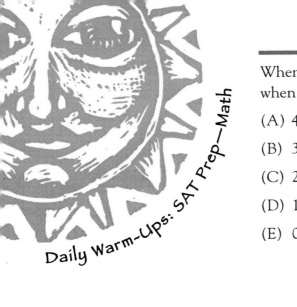

Daily Warm-Ups: SAT Prep—Math

Numbers and Operations

When r is divided by 5, the remainder is 3. What is the remainder when $3r$ is divided by 5?

(A) 4

(B) 3

(C) 2

(D) 1

(E) 0

  Ⓐ Ⓑ Ⓒ Ⓓ Ⓔ

17

Numbers and Operations

A store is offering a sale on T-shirts. If the T-shirts come in 4 different styles and 9 different colors, how many different T-shirts are on sale?

(A) 5

(B) 13

(C) 36

(D) 81

(E) 216

18

Ⓐ Ⓑ Ⓒ Ⓓ Ⓔ

Numbers and Operations

Monique sold 50 magazine subscriptions in the first week of her school fund-raiser. In the second week, she sold 30% fewer subscriptions. How many subscriptions did Monique sell in this two-week period?

(A) 20

(B) 35

(C) 70

(D) 80

(E) 85

(A) (B) (C) (D) (E)

19

Numbers and Operations

The average of 7 consecutive integers is 42. What is the greatest of these integers?

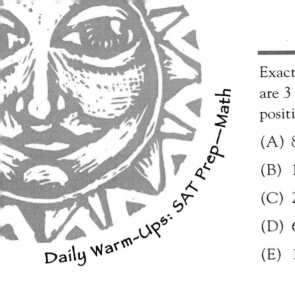

Numbers and Operations

Exactly 5 students are interested in serving as class officers. There are 3 positions open, and no student can fill more than one position. How many different assignments of positions are possible?

(A) 8

(B) 15

(C) 20

(D) 60

(E) 120

    (A) (B) (C) (D) (E)

21

Numbers and Operations

Several people are sitting in a row of seats in a movie theater. Starting at the right end of the row, Maureen is counted as the 9th person. Starting at the left end, she is counted as the 15th person. How many people are seated in the row?

(A) 6

(B) 7

(C) 22

(D) 23

(E) 24

22

© 2006 Walch Publishing

Ⓐ Ⓑ Ⓒ Ⓓ Ⓔ

Numbers and Operations

Alfie drove to school at an average rate of 30 miles per hour and returned home later in the day at the same rate of speed. If each leg of his trip took 15 minutes, how many miles did Alfie drive in total?

(A) 7.5

(B) 15

(C) 22.5

(D) 30

(E) 45

23

If *a* is the greatest common factor of 18 and 30, and *b* is the least common multiple of 8 and 12, what is the value of *ab*?

(A) 96

(B) 144

(C) 192

(D) 288

(E) 576

24

Ⓐ Ⓑ Ⓒ Ⓓ Ⓔ

Numbers and Operations

A factory produced 10,000 trading cards in 1 hour. If 2 percent of these cards are defective, how many cards are <u>not</u> defective?

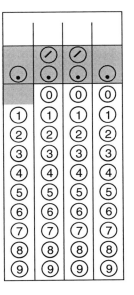

Numbers and Operations

Set A = {21, 29, 30, 37, 42, 47, 51, 53}

What fraction of the numbers in Set A are prime?

(A) $\dfrac{1}{4}$

(B) $\dfrac{3}{8}$

(C) $\dfrac{1}{2}$

(D) $\dfrac{5}{8}$

(E) $\dfrac{3}{4}$

26

Ⓐ Ⓑ Ⓒ Ⓓ Ⓔ

Numbers and Operations

The same number of men and women are on a train car when it leaves the terminal station. At the first stop, 10 men get on the car and no one gets off. After this stop, there are twice as many men as women on the train car. How many people are on this train car?

(A) 10

(B) 15

(C) 20

(D) 30

(E) 40

    Ⓔ

27

Numbers and Operations

A number x is decreased by 10%, and then this new number is increased by 20%. What is the final number, in terms of x?

(A) $0.7x$

(B) $1.08x$

(C) $1.1x$

(D) $1.15x$

(E) $1.2x$

28

(A) (B) (C) (D) (E)

Numbers and Operations

5, 12, 33, 96, . . .

The first term in the sequence above is 5, and each term after the first is determined by subtracting *a* from the preceding term and then multiplying by *b*. What is the value of *a*?

(A) 1

(B) 2

(C) 3

(D) 4

(E) 5

Daily Warm-Ups: SAT Prep—Math

Numbers and Operations

A team has won 50% of the 10 games it played in the first half of the season. The team would like to win 70% of the total games for the season. What percent of the remaining 10 games must it win in order to do so? (Disregard the % sign when gridding your answer. For example, if the answer is 10%, mark 10 or .1 on the grid.)

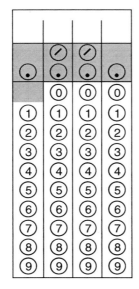

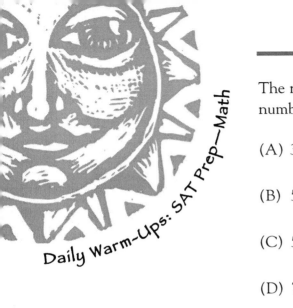

The reciprocal of a number is $\dfrac{5}{37}$. Which of the following is the number?

(A) $3\dfrac{7}{5}$

(B) 5.37

(C) $5\dfrac{3}{7}$

(D) $7\dfrac{2}{5}$

(E) 37.5

Ⓐ Ⓑ Ⓒ Ⓓ Ⓔ

If $40\% < a < \dfrac{5}{8}$, which of the following could be a?

(A) 0.25

(B) 0.50

(C) 0.65

(D) 0.70

(E) 0.85

Ⓐ Ⓑ Ⓒ Ⓓ Ⓔ

Numbers and Operations

Harry needs to select a number that is a multiple of 2, 3, and 5. Which of the following numbers could Harry select?

I. 650

II. 480

III. 1200

(A) I only

(B) I and II only

(C) II only

(D) II and III only

(E) I, II, and III

33

Daily Warm-Ups: SAT Prep—Math

(A) (B) (C) (D) (E)

Numbers and Operations

If 25% of x is 50, what is 50% of $4x$?

(A) 25

(B) 100

(C) 200

(D) 400

(E) 800

34

Numbers and Operations

If n is a number between 1500 and 2500 that can be expressed as the product of three consecutive two-digit integers, what is one possible value of n?

35

Numbers and Operations

$$x = 10{,}412$$

The units digit of x is increased by 3, its tens digit is increased by 5, and its thousands digit is increased by 2 to create the number y. How much greater is y than x?

(A) 352

(B) 253

(C) 2053

(D) 12,465

(E) 20,053

Daily Warm-Ups: SAT Prep—Math

36

© 2006 Walch Publishing

Ⓐ Ⓑ Ⓒ Ⓓ Ⓔ

Numbers and Operations

Daisy typed 3 pages in 5 minutes. At this rate, how many pages would she type in 8 hours?

(A) $4\frac{4}{5}$

(B) $13\frac{1}{3}$

(C) 120

(D) 288

(E) 800

$$4, 12, 36, 108, \ldots$$

The first term in the sequence above is 4, and each term after the first is determined by multiplying the preceding term by 3. What is the value of the 50th term of the sequence?

(A) 4×50

(B) $4 + (3 \times 50)$

(C) 50×3^4

(D) 4×3^{49}

(E) 4×3^{50}

38

Ⓐ Ⓑ Ⓒ Ⓓ Ⓔ

Numbers and Operations

Set X = {1, 4, 5, 6, 9, 12, 20}

The intersection of sets X and Y is {4, 9, 20}. Which of the following could be set Y?

(A) {3, 4, 9, 15, 20}

(B) {1, 5, 6, 12}

(C) {1, 4, 7, 9, 13, 20}

(D) {9, 14, 20}

(E) {2, 4, 9, 11, 21}

(A) (B) (C) (D) (E)

39

Numbers and Operations

A club is planning a trip to an upcoming concert. In order to pay the group rate, they must reserve entire rows of seats. Each row in the concert hall contains 16 seats. If there are 300 members in the group, how many rows must they reserve?

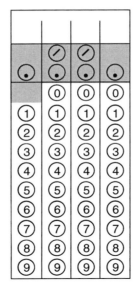

Daily Warm-Ups: SAT Prep—Math

Numbers and Operations

If purchased individually, a can of tuna costs $0.79. A multipack of 12 cans costs $7.80. How much money would be saved by purchasing multipacks instead of 60 individual cans of tuna?

(A) $0.14

(B) $1.68

(C) $7.01

(D) $8.40

(E) $100.80

41

Ⓐ Ⓑ Ⓒ Ⓓ Ⓔ

Numbers and Operations

A fruit punch is made by combining 1 quart of grape juice, 3 quarts of orange juice, 2 quarts of pineapple juice, and 2 quarts of ginger ale. What percent of the final mixture is pineapple juice?

(A) 12.5%

(B) 25%

(C) 33.$\bar{3}$%

(D) 37.5%

(E) 50%

42

   Ⓐ Ⓑ Ⓒ Ⓓ Ⓔ

Numbers and Operations

If A is the set of even integers less than 10, and B is the set of multiples of 3 less than 20, what is the union of sets A and B?

(A) {2, 4, 6, 8}

(B) {3, 6, 9, 12, 15, 18}

(C) {6}

(D) {2, 3, 4, 6, 8, 9, 12, 15, 18}

(E) {2, 4, 6, 8, 10, 12, 14, 16, 18}

(A) (B) (C) (D) (E)

43

Maggie collects marbles. She has one jar filled with 50 marbles and 4 jars each filled with 30 marbles. How many marbles does Maggie have in total?

(A) 80

(B) 150

(C) 170

(D) 230

(E) 250

44

  Ⓐ Ⓑ Ⓒ Ⓓ Ⓔ

Numbers and Operations

What is the smallest four-digit integer that has a factor of 7?

45

© 2006 Walch Publishing

The denominator of a certain fraction is 9 more than its numerator. If the fraction is equal to $\frac{2}{5}$, what is the denominator of this fraction?

(A) 4

(B) 6

(C) 10

(D) 15

(E) 20

Daily Warm-Ups: SAT Prep—Math

46

Ⓐ Ⓑ Ⓒ Ⓓ Ⓔ

Algebra and Functions

If a and b are inversely proportional, and $a = 7$ when $b = 6$, what is a when b is 21?

(A) 2

(B) 13

(C) 18

(D) 20

(E) 42

47

(A) (B) (C) (D) (E)

© 2006 Walch Publishing

Algebra and Functions

If $6a + 3b = 6$, what is the value of $2a + b$?

(A) 1

(B) 2

(C) 3

(D) 12

(E) 18

48

(A) (B) (C) (D) (E)

Algebra and Functions

If $xyz < 0$ and $xy > 0$, which of the following MUST be true?

(A) $x > 0$

(B) $y > 0$

(C) $z < 0$

(D) $xz < 0$

(E) $yz > 0$

Ⓐ Ⓑ Ⓒ Ⓓ Ⓔ

49

If $a + 2b = 21$ and $3a = b$, what is the value of b?

Algebra and Functions

If $3^{3x} = 9^{x+1}$, what is the value of x?

(A) 0

(B) 1

(C) 2

(D) 3

(E) 9

Ⓐ Ⓑ Ⓒ Ⓓ Ⓔ

51

Algebra and Functions

What is the value of $3y^2 + 2y$, when $y = -1$?

(A) -5

(B) -3

(C) -1

(D) 1

(E) 5

52

 Ⓐ Ⓑ Ⓒ Ⓓ Ⓔ

Algebra and Functions

If $x^2 + y^2 = 85$ and $(x - y)^2 = 49$, what is the value of xy?

(A) 18

(B) 36

(C) 54

(D) 81

(E) 134

Ⓐ Ⓑ Ⓒ Ⓓ Ⓔ

Algebra and Functions

If $4x + 3 > -13$, all of the following can be x EXCEPT

(A) -5

(B) -3

(C) 0

(D) 2

(E) 4

54

(A) (B) (C) (D) (E)

Daily Warm-Ups: SAT Prep—Math

If $4^{n-1} = 8^2$, then $n =$

55

Algebra and Functions

Five less than twice a number is 11. What is the number?

(A) –6

(B) –3

(C) 3

(D) 8

(E) 16

56

 Ⓐ Ⓑ Ⓒ Ⓓ Ⓔ

If $f(x) = 5x^2 - 8x + 3$, then $f(-3) =$

(A) −63

(B) −36

(C) 24

(D) 48

(E) 72

Daily Warm-Ups: SAT Prep—Math

(A) (B) (C) (D) (E)

57

Algebra and Functions

$$\frac{x^2 - y^2}{3x - 3y} = ?$$

(A) $\dfrac{1}{3}$

(B) 3

(C) $\dfrac{3}{x + y}$

(D) $\dfrac{x - y}{3}$

(E) $\dfrac{x + y}{3}$

58

Ⓐ   Ⓑ   Ⓒ   Ⓓ   Ⓔ

Algebra and Functions

How much less than $x + 9$ is $x - 3$?

(A) 3

(B) 6

(C) 9

(D) 12

(E) 27

Ⓐ Ⓑ Ⓒ Ⓓ Ⓔ

Algebra and Functions

Let the operation @ be defined by $m@n = nm + n$ for all numbers m and n. If $5@7 = x@3$, what is the value of x?

Algebra and Functions

If $3x - 2y = 8$ and $x - 4y = 3$, what is the value of $x + y$?

(A) $-\dfrac{1}{10}$

(B) $2\dfrac{1}{2}$

(C) $2\dfrac{3}{5}$

(D) $2\dfrac{7}{10}$

(E) 5

Ⓐ Ⓑ Ⓒ Ⓓ Ⓔ

Algebra and Functions

If $5n + 8 = 2n + 20$, then $n =$

(A) 3

(B) 4

(C) 6

(D) 7

(E) 12

62

(A) (B) (C) (D) (E)

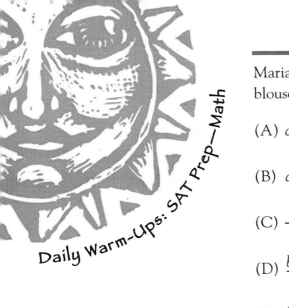

Algebra and Functions

Mariana buys b blouses for d dollars. How many dollars do $b - 1$ blouses cost?

(A) $d - 1$

(B) $d(b - 1)$

(C) $\dfrac{d}{b - 1}$

(D) $\dfrac{b(b - 1)}{d}$

(E) $\dfrac{d(b - 1)}{b}$

63

Algebra and Functions

$$(a^3)^5 =$$

(A) $a^{\frac{3}{5}}$

(B) a^2

(C) a^8

(D) a^{15}

(E) a^{243}

64

Ⓐ Ⓑ Ⓒ Ⓓ Ⓔ

If $n^4 = 256$, what is the value of $5n^4$?

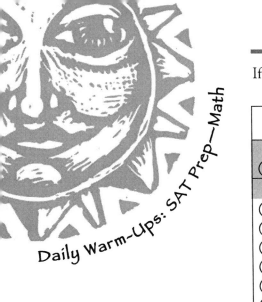

65

© 2006 Walch Publishing

Algebra and Functions

If $a - 4b = c$, what is b in terms of a and c?

(A) $\dfrac{ac}{4}$

(B) $\dfrac{c - a}{4}$

(C) $\dfrac{a - c}{4}$

(D) $-\dfrac{c}{4a}$

(E) $-\dfrac{a}{4c}$

66

Ⓐ Ⓑ Ⓒ Ⓓ Ⓔ

Which of the following is a root of the equation $x^2 + x - 6 = 0$?

(A) -6

(B) -3

(C) -2

(D) 0

(E) 1

Ⓐ Ⓑ Ⓒ Ⓓ Ⓔ

Algebra and Functions

$$8^{\frac{2}{3}} =$$

(A) 2^4

(B) 4^3

(C) $16^{\frac{1}{2}}$

(D) $32^{\frac{3}{2}}$

(E) 64^0

68

Ⓐ Ⓑ Ⓒ Ⓓ Ⓔ

Daily Warm-Ups: SAT Prep—Math

If $|x + 5| + 3 = 11$, the value of x could be

   I.   −13

  II.   3

 III. 19

(A) I only

(B) II only

(C) I and II only

(D) II and III only

(E) I, II, and III

69

If $3(4x - 3) = 2x + 1$, what is the value of x?

Algebra and Functions

For which of the following functions is $f(1) > f(2)$?

(A) $f(x) = 2x - 1$

(B) $f(x) = x^2 + x$

(C) $f(x) = \dfrac{x}{x + 1}$

(D) $f(x) = 3^x$

(E) $f(x) = 3 - x^2$

Ⓐ   Ⓑ   Ⓒ   Ⓓ   Ⓔ

71

Algebra and Functions

Which of the following describes the solution set to the inequality $5 + x \leq 2x - 3$?

(A) $x \geq 8$

(B) $x \leq 8$

(C) $x \geq 2$

(D) $x \leq 2$

(E) $x = 4$

72

Ⓐ Ⓑ Ⓒ Ⓓ Ⓔ

Algebra and Functions

If m and n are directly proportional, and $m = -2$ when $n = 6$, what is the value of m when $n = -12$?

(A) -36

(B) -6

(C) -3

(D) 2

(E) 4

Algebra and Functions

What is the domain of the function $f(x) = \sqrt{x-4}$?

(A) $x > 4$

(B) $x \geq 4$

(C) $x \geq 2$

(D) $x > -4$

(E) $x \leq 4$

74

(A) (B) (C) (D) (E)

Algebra and Functions

Let 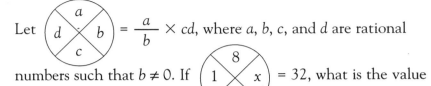 $= \dfrac{a}{b} \times cd$, where a, b, c, and d are rational numbers such that $b \neq 0$. If $\left(\begin{smallmatrix} 8 \\ 1 \;\;\; x \\ 2 \end{smallmatrix} \right) = 32$, what is the value of x?

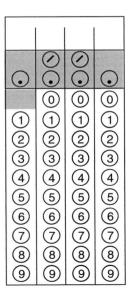

75

Algebra and Functions

If $y = mx + b$, what is m in terms of x, y, and b?

(A) $\dfrac{y - b}{x}$

(B) $\dfrac{x + b}{y}$

(C) $\dfrac{b - y}{x}$

(D) $\dfrac{x - y}{b}$

(E) $\dfrac{x - b}{y}$

76

(A) (B) (C) (D) (E)

Algebra and Functions

x	0	1	2	3
$f(x)$	3	2	1	0

The table above gives values for the function f for selected values of x. Which of the following defines f?

(A) $f(x) = x + 3$

(B) $f(x) = -x + 3$

(C) $f(x) = -x$

(D) $f(x) = x + 1$

(E) $f(x) = x - 1$

77

Daily Warm-Ups: SAT Prep—Math

Algebra and Functions

$$\frac{2x^2 - 3x - 2}{6x + 3} =$$

(A) $\dfrac{x-2}{3}$

(B) $\dfrac{2^2 x}{3x+1}$

(C) $\dfrac{2x+1}{x-2}$

(D) $\dfrac{2x+1}{3x}$

(E) $\dfrac{x}{3-x}$

78

Ⓐ Ⓑ Ⓒ Ⓓ Ⓔ

Two years ago, Sophie was *x* years old. How old will she be in *y* years?

(A) $x + y$

(B) $x + y - 2$

(C) $x - y + 2$

(D) $x + y + 2$

(E) $x - y - 2$

79

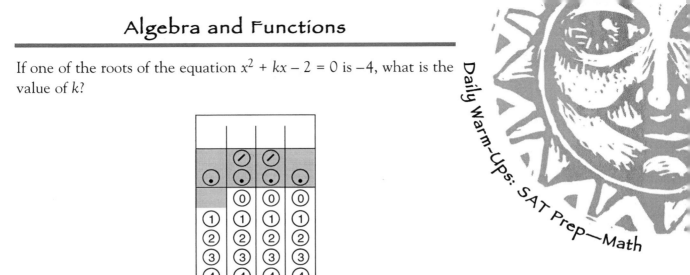

Algebra and Functions

If one of the roots of the equation $x^2 + kx - 2 = 0$ is -4, what is the value of k?

80

Algebra and Functions

Line 1 is defined by the equation $2y + 3x = 4$. What is the slope of line 1?

(A) $-\dfrac{3}{2}$

(B) $\dfrac{3}{2}$

(C) 2

(D) 3

(E) 4

Algebra and Functions

How many inches are there in f feet and y yards?

(A) $\dfrac{f}{12} + \dfrac{y}{36}$

(B) $24fy$

(C) $48(f + y)$

(D) $12f + y$

(E) $12(f + 3y)$

82

Ⓐ Ⓑ Ⓒ Ⓓ Ⓔ

Daily Warm-Ups: SAT Prep—Math

Algebra and Functions

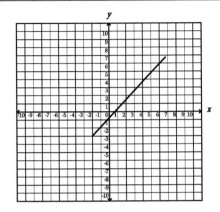

Which of the following statements about the linear function graphed above is true?

(A) It has a positive slope and a positive y-intercept.

(B) It has a positive slope and a negative y-intercept.

(C) It has a negative slope and a positive y-intercept.

(D) It has a negative slope and a negative y-intercept.

(E) Its slope is undefined.

83

© 2006 Walch Publishing

Algebra and Functions

A phone company charges x for the first 5 minutes of a phone call and y for each minute beyond that. If $z > x$, which of the following represents the cost of a phone call that is z minutes long?

(A) xyz

(B) $x + yz$

(C) $5x + yz$

(D) $x + y(z - 5)$

(E) $5x + y(z - 5)$

84

(A) (B) (C) (D) (E)

Algebra and Functions

An object thrown upwards is f feet from the ground after t seconds, where $f = 80t - 16t^2$. At 2 seconds, how many feet above the ground will the object be?

© 2006 Walch Publishing

Algebra and Functions

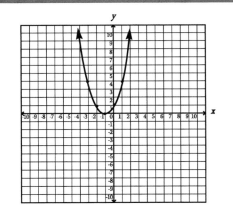

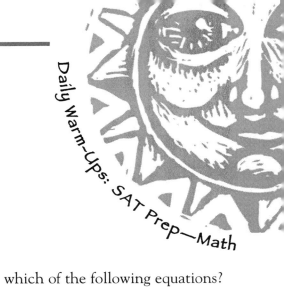

The parabola graphed above is represented by which of the following equations?

(A) $y = x^2 + 2x + 1$

(B) $y = -x^2 + 2x + 1$

(C) $y = x^2 + 2x - 3$

(D) $y = -x^2 + 2x - 3$

(E) $y = x^2 + x + 1$

Ⓐ Ⓑ Ⓒ Ⓓ Ⓔ

86

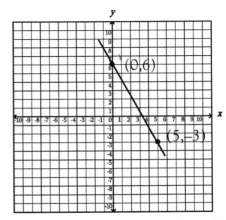

If the slope of the linear function graphed above is −1, what is the value of b?

(A) −5

(B) 1

(C) 2

(D) 3

(E) 8

Ⓐ Ⓑ Ⓒ Ⓓ Ⓔ

Algebra and Functions

Two lines intersect at the point $(0, 5)$. If the equation of one of these lines is $y - 2x = 5$, which of the following could be the equation of the second line?

(A) $2y + x = -4$

(B) $2y - x = 5$

(C) $3y - x = -15$

(D) $4y + x = 20$

(E) $y = x$

88

  (A) (B) (C) (D) (E)

Daily Warm-Ups: SAT Prep—Math

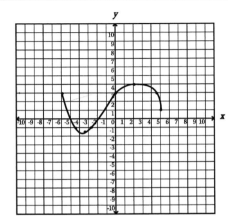

The graph of $y = f(x)$ is shown above. All of the following statements about this graph are true EXCEPT

(A) The domain of the function is the interval from $x = -6$ to $x = 5$.

(B) The range of the function is the interval from $y = -1.5$ to $y = 4$.

(C) $f(x) = 3$ for 3 values of x.

(D) $f(x) = -1$ for 1 value of x.

(E) There are 2 zeros for the function f.

    Ⓐ Ⓑ Ⓒ Ⓓ Ⓔ

89

© 2006 Walch Publishing

Algebra and Functions

If $2a + 3b = 29$ and $a = 2b - 3$, what is the value of b?

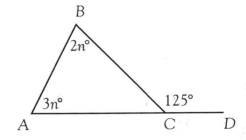

In the figure above, what is the measure of angle A?

(A) 25°

(B) 50°

(C) 55°

(D) 75°

(E) 100°

Geometry and Measurement

A square has an area of 36 square feet. Its length is increased by 4 feet and its width is decreased by 2 feet to create a rectangle. What is the area of this rectangle in square feet?

(A) 28

(B) 38

(C) 40

(D) 52

(E) 64

92

   Ⓐ Ⓑ Ⓒ Ⓓ Ⓔ

Geometry and Measurement

In right triangle XYZ, XZ = 12 inches, YZ = 13 inches, and XY = 5 inches. What is the area of the triangle in square inches?

(A) 30

(B) 32.5

(C) 60

(D) 65

(E) 78

93

© 2006 Walch Publishing

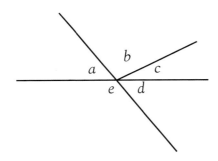

In the figure above, which angle is supplementary to angle *a*?

(A) *b*

(B) *c*

(C) *d*

(D) *e*

(E) None

Points A, B, and C all lie on line l such that $AB = 4CB$. If $CB = 2$ and $AC < AB$, what is AC?

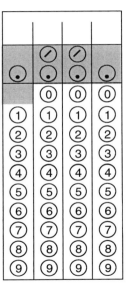

95

Geometry and Measurement

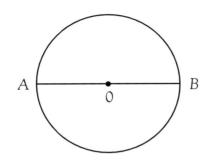

If $AB = 10$ units, what is the area of the circle, in square units?

(A) 5π

(B) 10π

(C) 25π

(D) 50π

(E) 100π

96

Ⓐ Ⓑ Ⓒ Ⓓ Ⓔ

Daily Warm-Ups: SAT Prep—Math

Geometry and Measurement

Line 1 is parallel to line 2. If the equation of line 1 is $3y - x = 6$, which of the following could be the equation of line 2?

(A) $y = x + 3$

(B) $y = -x + 2$

(C) $y = 3x$

(D) $y = -\dfrac{1}{3}x + 2$

(E) $y = \dfrac{1}{3}x$

Ⓐ Ⓑ Ⓒ Ⓓ Ⓔ

97

© 2006 Walch Publishing

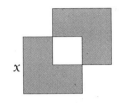

The figure above is formed by two identical squares overlapping such that a smaller square is created. Two corners of this smaller square lie on the centers of the larger squares, and two lie on the midpoints of sides of the larger squares. If the sides of the larger squares have length x units, what is the area of the shaded region in square units?

(A) $\dfrac{x^2}{4}$

(B) $\dfrac{3x^2}{2}$

(C) $\dfrac{7x^2}{4}$

(D) x^2

(E) $2x^2$

Ⓐ Ⓑ Ⓒ Ⓓ Ⓔ

Daily Warm-Ups: SAT Prep—Math

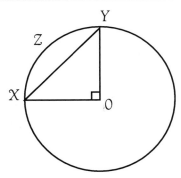

In the figure above, the arc XZY has length 5π and $\angle XOY$ is a right angle. What is the area of triangle XOY in square units?

(A) 25

(B) 50

(C) 100

(D) 200

(E) It cannot be determined from the information given.

(A)   (B)   (C)   (D)   (E)

Geometry and Measurement

Angle W is complementary to angle X. Angle X is supplementary to angle Y. Angle Y and angle Z are vertical angles. If angle Z contains 100°, how many degrees does angle W contain?

In isosceles triangle ABC, $AB = 5$ and $BC = 8$. Which of the following could be the perimeter of this triangle?

   I.   13

  II.   20

 III.   21

(A) I only

(B) II only

(C) III only

(D) I and II only

(E) II and III only

101

© 2006 Walch Publishing

Geometry and Measurement

A wooden cube has a volume of 64. If one third of its faces are painted green, what is the total area of the surface of the cube that is NOT green?

(A) 16

(B) 32

(C) 64

(D) 80

(E) 96

102

Ⓐ Ⓑ Ⓒ Ⓓ Ⓔ

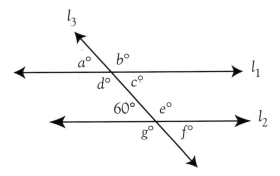

Parallel lines 1 and 2 are intersected by line 3. If $b = 2f$, what is the value of $a + c + d$?

(A) 150°

(B) 180°

(C) 240°

(D) 300°

(E) 360°

Geometry and Measurement

What is the area of the largest circle that can be drawn within a square of perimeter 36?

(A) 6π

(B) 9π

(C) 13.5π

(D) 20.25π

(E) 81π

104

Geometry and Measurement

In the triangle at the right, $MN = 7$ units and $NO = 5$ units. What could be the number of units in the perimeter of this triangle?

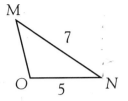

Geometry and Measurement

A rectangular prism has measurements of 3 by 8 by 10. Which of the following sets of measurements describes a rectangular prism with the same volume?

(A) 5 by 6 by 10

(B) 4 by 6 by 8

(C) 3 by 12 by 15

(D) 2 by 8 by 12

(E) 4 by 5 by 12

106

(A) (B) (C) (D) (E)

Line segment $\overline{XY}$ has endpoints X (–2, 5) and Y (4, y). If its midpoint is (1, 2), what is the value of y?

(A) –3

(B) –1

(C) $-\dfrac{1}{2}$

(D) $3\dfrac{1}{2}$

(E) 5

(A) (B) (C) (D) (E)

107

Geometry and Measurement

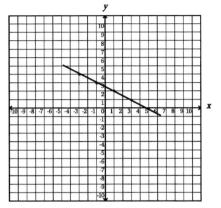

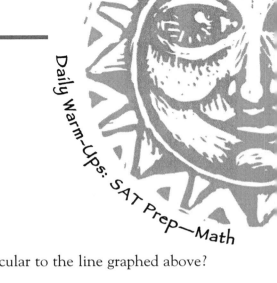

Which of the following equations represents a line perpendicular to the line graphed above?

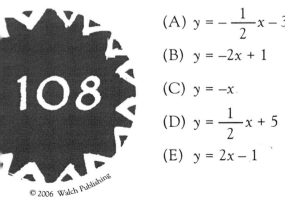

(A) $y = -\dfrac{1}{2}x - 3$

(B) $y = -2x + 1$

(C) $y = -x$

(D) $y = \dfrac{1}{2}x + 5$

(E) $y = 2x - 1$

Ⓐ Ⓑ Ⓒ Ⓓ Ⓔ

Daily Warm-Ups: SAT Prep—Math

Geometry and Measurement

Each of the following describes the measurements of a right circular cylinder. Which of these cylinders has the greatest volume?

(A) $r = 3, h = 10$

(B) $r = 4, h = 6$

(C) $r = 5, h = 3$

(D) $r = 2, h = 15$

(E) $r = 1, h = 50$

   Ⓐ Ⓑ Ⓒ Ⓓ Ⓔ

109

Geometry and Measurement

The three sides of triangle ABC have different integer side lengths. If $AB = 8$ and $BC = 13$, how many different values are possible for the length of AC?

110

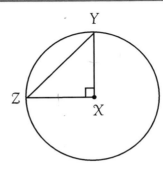

In the figure above, X is the center of a circle whose circumference is 64π. Y and Z are points on the circle. If angle YXZ is a right angle, what is the length of line segment $\overline{YZ}$?

(A) 8

(B) $8\sqrt{2}$

(C) 16

(D) 32

(E) $32\sqrt{2}$

Ⓐ   Ⓑ   Ⓒ   Ⓓ   Ⓔ

Geometry and Measurement

A certain regular polygon contains a total of 180x degrees. In terms of x, how many sides does this polygon have?

(A) $x - 2$

(B) $x - 1$

(C) $x + 1$

(D) $x + 2$

(E) $\dfrac{180}{x}$

112

Ⓐ Ⓑ Ⓒ Ⓓ Ⓔ

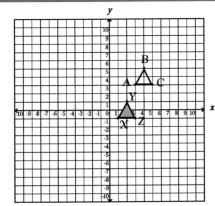

In the figure above, triangle ABC has been translated to produce the shaded triangle XYZ. Which of the following best describes the translation of triangle ABC?

(A) a translation 4 units up

(B) a translation 2 units left

(C) a translation 4 units down and 2 units left

(D) a translation 4 units down and 2 units right

(E) a translation 2 units up and 4 units down

Geometry and Measurement

Line segment *RS* has endpoints (–2, 6) and (–7, –3). What is *RS*?

(A)   2

(B) $3\sqrt{10}$

(C) $4\sqrt{5}$

(D) $\sqrt{106}$

(E)   14

114

Ⓐ Ⓑ Ⓒ Ⓓ Ⓔ

Geometry and Measurement

Quin pushes a circular wheel with a radius of 8 feet. He wants to see how far he can make it roll before it stops and falls over. If the wheel rolls 50π feet before it stops, how many revolutions did it make?

115

Geometry and Measurement

Two sides of a right triangle have lengths 6 and 8. What is the area of this triangle?

(A) $6\sqrt{7}$

(B) 24

(C) $12\sqrt{7}$

(D) 48

(E) It cannot be determined from the information given.

116

Ⓐ Ⓑ Ⓒ Ⓓ Ⓔ

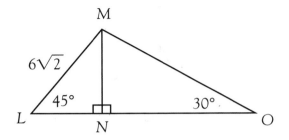

In the figure above, $LM = 6\sqrt{2}$, angle $L = 45°$, and angle $O = 30°$.
What is the area of triangle LMO?

(A) 18

(B) $18\sqrt{3}$

(C) $18(1 + \sqrt{3})$

(D) $6(3 + \sqrt{2} + \sqrt{3})$

(E) $48\sqrt{2}$

Geometry and Measurement

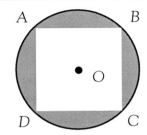

In the figure above, square $ABCD$ is inscribed within circle O. If the perimeter of the square is 32, what is the ratio of the shaded area to the unshaded area?

(A) $\dfrac{\pi - 1}{1}$

(B) $\dfrac{\pi - 2}{2}$

(C) $\dfrac{2\pi - 1}{1}$

(D) $\dfrac{\pi}{2}$

(E) $\dfrac{\pi}{4}$

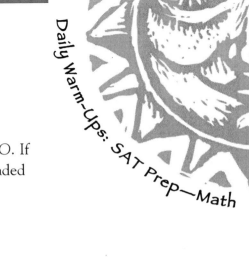

118

Ⓐ Ⓑ Ⓒ Ⓓ Ⓔ

Daily Warm-Ups: SAT Prep—Math

Geometry and Measurement

The lengths of two sides of an isosceles triangle are 8 and 12. What is the greatest possible value for the perimeter of the triangle?

(A) 20

(B) 28

(C) 32

(D) 36

(E) 96

(A) (B) (C) (D) (E)

119

© 2006 Walch Publishing

Geometry and Measurement

Eliana draws a polygon that contains a total of n degrees. If $500 < n < 1,000$, what is a possible value for the number of sides in this polygon?

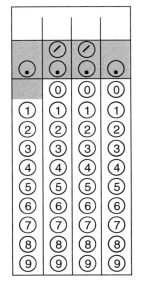

120

Geometry and Measurement

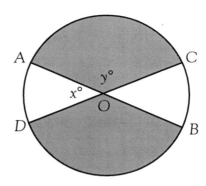

In the figure above, diameters AB and CD intersect at O, the center of the circle. If $y = 5x$, what fractional part of the circle is shaded?

(A) $\dfrac{1}{12}$ (B) $\dfrac{1}{6}$ (C) $\dfrac{1}{5}$ (D) $\dfrac{4}{5}$ (E) $\dfrac{5}{6}$

Ⓐ Ⓑ Ⓒ Ⓓ Ⓔ

121

Geometry and Measurement

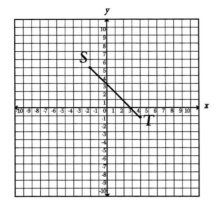

What is the length of ST?

(A) $2\sqrt{5}$

(B) 6

(C) $6\sqrt{2}$

(D) $6\sqrt{3}$

(E) 12

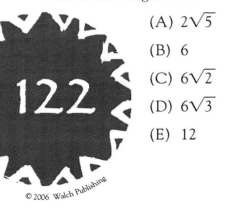

122

Triangle 1 is similar to triangle 2. The height of triangle 1 is 3, and the height of triangle 2 is 6. If the area of triangle 2 is 12, what is the area of triangle 1?

(A) 3

(B) 6

(C) 9

(D) 24

(E) 48

123

Ⓐ Ⓑ Ⓒ Ⓓ Ⓔ

© 2006 Walch Publishing

Geometry and Measurement

Which of the following represents a reflection?

(A)

(B)

(C)

(D)

(E)

124

    Ⓔ

A B C D E

Geometry and Measurement

A square has an area of a^2. Its width is increased by n and its length is decreased by 2 to create a rectangle. If the area of the rectangle is $a^2 + a - 6$, what is the value of n?

In triangle XYZ, XY = YZ = 1. Which of the following MUST be true?

I. $\angle Y = 60°$

II. $\angle Z = \angle X$

III. $XZ = \sqrt{2}$

(A) II only

(B) I and II only

(C) II and III only

(D) I and III only

(E) I, II, and III

Daily Warm-Ups: SAT Prep—Math

126

Geometry and Measurement

Square MNOP is graphed on the coordinate axes and has an area of 25. If point M has coordinates $(2, 3)$, which of the following could NOT be the coordinates of another corner of MNOP?

(A) $(5, 8)$

(B) $(2, 8)$

(C) $(7, -2)$

(D) $(7, 3)$

(E) $(-3, -2)$

127

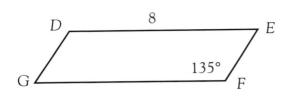

In parallelogram *DEFG*, *DE* = 8 and ∠*F* = 135°. If the area of the parallelogram is 16, what is *DG*?

(A) 2

(B) 4

(C) 6

(D) $2\sqrt{2}$

(E) $4\sqrt{2}$

Daily Warm-Ups: SAT Prep—Math

Ⓐ Ⓑ Ⓒ Ⓓ Ⓔ

Geometry and Measurement

In a certain triangle, the angles have measures of $3n°$, $4n°$, and $5n°$. What is the measure of the largest angle in this triangle?

(A) 60°

(B) 75°

(C) 90°

(D) 105°

(E) 135°

129

    (A) (B) (C) (D) (E)

© 2006 Walch Publishing

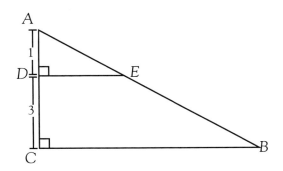

In the figure above, $AD = 1$ and $DC = 3$. If the area of triangle ABC is 24, what is DE?

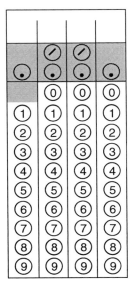

Daily Warm-Ups: SAT Prep—Math

Geometry and Measurement

All the following sets of triangle side lengths satisfy the Pythagorean theorem EXCEPT

(A) 9, 12, 15

(B) $5\sqrt{2}$, $5\sqrt{2}$, 10

(C) 5, $\sqrt{119}$, 12

(D) $2\sqrt{3}$, 6, $4\sqrt{3}$

(E) 13, 14, 15

  Ⓐ Ⓑ Ⓒ Ⓓ Ⓔ

© 2006 Walch Publishing

Geometry and Measurement

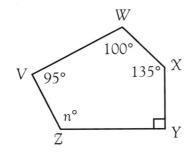

In the figure above, what is the value of n?

(A) 30

(B) 90

(C) 105

(D) 120

(E) 130

132

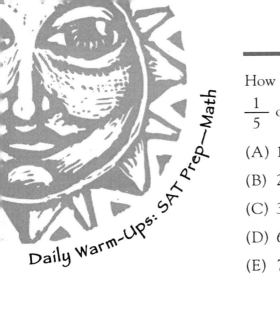

How many fewer degrees of arc are there in $\frac{1}{6}$ of a circle than in $\frac{1}{5}$ of a circle?

(A) 12°

(B) 24°

(C) 36°

(D) 60°

(E) 72°

Daily Warm-Ups: SAT Prep—Math

The figure at the right represents a rectangular prism with height 12, length 3, and width 4. What is the distance between points B and H?

(A) $\sqrt{153}$

(B) $4\sqrt{10}$

(C) 13

(D) 15

(E) It cannot be determined from the information given.

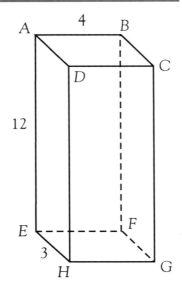

134

Ⓐ Ⓑ Ⓒ Ⓓ Ⓔ

Geometry and Measurement

The diameter of circle A is 9 times as great as the diameter of circle B. The area of circle A is how many times as great as the area of circle B?

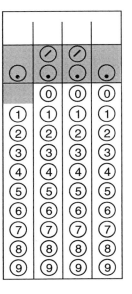

© 2006 Walch Publishing

Data Analysis, Statistics, and Probability

What is the average of x, $x + 1$, $x - 1$, and 0?

(A) $\dfrac{x}{3}$

(B) $\dfrac{3x}{4}$

(C) x

(D) $x + \dfrac{1}{4}$

(E) $2x$

136

  Ⓐ Ⓑ Ⓒ Ⓓ Ⓔ

Data Analysis, Statistics, and Probability

There are 20 marbles in a jar. If the probability of selecting a blue marble is $\frac{2}{5}$, how many marbles in the jar are NOT blue?

(A) 2

(B) 4

(C) 8

(D) 12

(E) 16

137

(A) (B) (C) (D) (E)

Data Analysis, Statistics, and Probability

The median of a group of numbers is 17. Which of the following could be this group of numbers?

(A) 12, 17, 18, 14

(B) 17, 19, 17, 25, 1, 20

(C) 57, 17, 37

(D) 48, 1, 2

(E) 1, 17, 100, 8, 59

138

   Ⓓ Ⓔ

Data Analysis, Statistics, and Probability

Isabel is making a pictograph to represent the number of days that it rained each month from March through July. On her graph, an umbrella icon represents 2 rainy days. If it rained for 9 days in April, how should this be represented on Isabel's pictograph?

(A)

(B)

(C)

(D)

(E)

139

Data Analysis, Statistics, and Probability

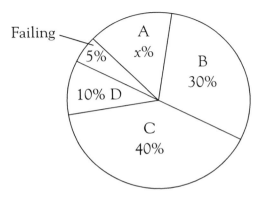

The graph above shows the scores of the junior class on a recent school-wide exam. If 96 students received a C, how many students received an A?

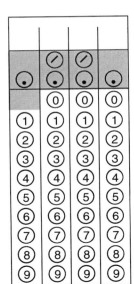

140

Data Analysis, Statistics, and Probability

The ratio of black socks to white socks in a drawer is 1:4. What is the probability that a sock removed from the drawer at random will be white?

(A) $\dfrac{1}{5}$

(B) $\dfrac{1}{4}$

(C) $\dfrac{3}{4}$

(D) $\dfrac{4}{5}$

(E) It cannot be determined from the information given.

141

Data Analysis, Statistics, and Probability

Matt's average for his first five math tests is 88. What must he score on his sixth test to raise his average to 90?

(A) 89

(B) 90

(C) 92

(D) 98

(E) 100

142

Ⓐ Ⓑ Ⓒ Ⓓ Ⓔ

Data Analysis, Statistics, and Probability

Month	Hours Worked
July	60
August	78
September	40
October	48
November	50
December	60

The chart above shows the number of hours that Sarah worked at her part-time job each month from July through September. In what month did she work 30% more hours than she worked the month before?

(A) August

(B) September

(C) October

(D) November

(E) December

Daily Warm-Ups: SAT Prep—Math

143

Data Analysis, Statistics, and Probability

Mike scored an average of 3 goals per game this soccer season. If he played in 24 games this season, what is the total number of goals he scored?

(A) 8

(B) 21

(C) 27

(D) 48

(E) 72

144

Ⓐ Ⓑ Ⓒ Ⓓ Ⓔ

Data Analysis, Statistics, and Probability

The average of 9 consecutive odd integers is 63. What is the greatest of these integers?

A service club is holding a lottery drawing for its members. Each member writes his or her name on 5 slips of paper, which are then folded and placed in a large box. One slip of paper will be drawn for the first prize. After it is removed from the box, a second slip of paper will be removed for the grand prize. If there are 22 members in the club, what is the probability that the same person will win both prizes?

(A) $\dfrac{5}{22}$

(B) $\dfrac{10}{231}$

(C) $\dfrac{1}{484}$

(D) $\dfrac{2}{1199}$

(E) $\dfrac{1}{605}$

146

Ⓐ Ⓑ Ⓒ Ⓓ Ⓔ

Data Analysis, Statistics, and Probability

Set A = {1, 5, 8, 5, 8, 3, 5, 3, 8, 2, 1, 5, x}

In order for 5 to be the only mode for Set A, x could be all of the following EXCEPT

(A) 1

(B) 2

(C) 3

(D) 5

(E) 8

Data Analysis, Statistics, and Probability

Jose made a total of 15.75 pounds of potato salad for her family picnic. She divided the potato salad among 7 plastic containers to transport it. What is the average amount of salad, in pounds, in each of these containers?

(A) 2.25

(B) 3.15

(C) 5.25

(D) 8.75

(E) 11.375

148

(A) (B) (C) (D) (E)

Data Analysis, Statistics, and Probability

Number of Soft Drinks	Total Price
1	$0.95
6 pack	$4.79
Case of 24	$15.99

According to the information in the table above, what would be the least amount of money needed, in dollars, to purchase exactly 41 soft drinks?

(A) $21.73

(B) $30.32

(C) $31.98

(D) $38.32

(E) $38.95

149

The average of five numbers is 17. When a sixth number is added to the set, the new average is 15. What is the sixth number?

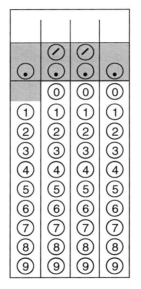

150

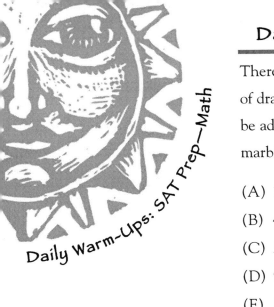

Data Analysis, Statistics, and Probability

There are red, green, and blue marbles in a bag. The probability of drawing a green marble is $\frac{3}{7}$. How many green marbles must be added to the bag to make the probability of drawing a green marble $\frac{1}{2}$?

(A) 1

(B) 4

(C) 5

(D) 9

(E) It cannot be determined from the information given.

151

© 2006 Walch Publishing

Ⓐ Ⓑ Ⓒ Ⓓ Ⓔ

Data Analysis, Statistics, and Probability

Mr. Sanders needs to purchase a total of 8 sleeping bags for his family's upcoming camping trip. He buys 3 secondhand from a neighbor for a total of $30. He buys one on-line for $54. He buys the remaining 4 at a chain store for $40 each. What is the average price that Mr. Sanders paid for a sleeping bag?

(A) $30.50

(B) $32.00

(C) $34.75

(D) $38.00

(E) $38.66

152

(A) (B) (C) (D) (E)

Data Analysis, Statistics, and Probability

The average of five numbers is 9. If the average of two of those number is $7\frac{1}{2}$, what is the average of the other three numbers?

(A) $\frac{1}{2}$

(B) $1\frac{1}{2}$

(C) 10

(D) $12\frac{1}{2}$

(E) 15

153

Data Analysis, Statistics, and Probability

Kecia has a standard deck of playing cards that contains 13 cards of each of the 4 suits. She removes 4 cards from the deck. If the probability of selecting a card with the suit diamonds is now $\frac{1}{4}$, how many of the cards removed were diamonds?

    (A) 0

    (B) 1

    (C) 2

    (D) 3

    (E) 4

154

Ⓐ Ⓑ Ⓒ Ⓓ Ⓔ

Data Analysis, Statistics, and Probability

Anya flips a quarter 4 times in a row. What is the probability that the quarter landed on tails all 4 times?

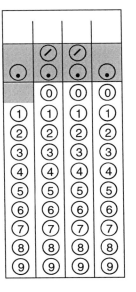

Data Analysis, Statistics, and Probability

Charitable donations to a certain group in 2003 totaled $2,125,000. If the average contribution was $425, what was the total number of contributions made?

(A) 50

(B) 250

(C) 500

(D) 2500

(E) 5000

156

Ⓐ Ⓑ Ⓒ Ⓓ Ⓔ

Data Analysis, Statistics, and Probability

Will buys raffle tickets consecutively numbered from 125 to 150. If a total of 750 tickets were sold, what is the probability that Will will win the raffle?

(A) $\dfrac{1}{750}$

(B) $\dfrac{1}{30}$

(C) $\dfrac{13}{375}$

(D) $\dfrac{1}{6}$

(E) $\dfrac{1}{5}$

Ⓐ Ⓑ Ⓒ Ⓓ Ⓔ

157

Data Analysis, Statistics, and Probability

Sophie drives from home to her grandmother's house at a rate of 60 miles per hour. When she drives back home again, there is traffic, so her rate of speed is 40 miles per hour for the return trip. What is her average rate of speed for the entire trip?

(A) 48

(B) 50

(C) 52

(D) 55

(E) It cannot be determined from the information given.

158

© 2006 Walch Publishing

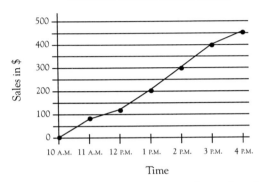

School Concession Sales

The graph above shows the sales at the school concession stand on the day of a football game. The greatest percent increase in sales occurred during which time period?

(A) 11 A.M. – 12 P.M.

(B) 12 P.M. – 1 P.M.

(C) 1 P.M. – 2 P.M.

(D) 2 P.M. – 3 P.M.

(E) 3 P.M. – 4 P.M.

Tatiana is playing a board game. In order to win on her next turn, she must roll a 5 or greater on a standard die. What is the probability that Tatiana will win on her next turn?

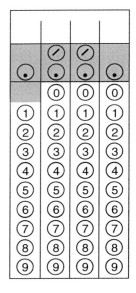

160

Data Analysis, Statistics, and Probability

Day	Number of Balloons Sold
Monday	♀ ♀ ♀ ♀
Tuesday	♀ ♀ ♀ ♀ ♀ (
Wednesday	♀ ♀ ♀ ♀ ♀ ♀
Thursday	♀ ♀ ♀ ♀ ♀
Friday	♀ ♀ ♀ ♀ ♀ ♀ ♀
Saturday	♀ ♀ ♀ ♀ ♀ ♀ ♀ ♀ ♀ ♀ ♀
Sunday	♀ ♀ ♀ ♀ ♀ ♀ ♀ ♀ ♀ ♀ ♀ ♀

♀ = 4 balloons

A street vendor sells balloons outside the entrance to the park. He normally works Monday through Saturday, taking Sundays off. For one week he works all seven days and tracks the number of balloons he sells each day. These data are compiled into the pictograph shown above. According to this pictograph, how many more balloons per week would the vendor sell if he took Monday off and worked on Sunday instead?

(A) 8   (B) 12   (C) 32   (D) 48   (E) 52

161

Data Analysis, Statistics, and Probability

A spinner is divided into 8 equal sections. In order for the probability of landing on blue in one spin to be 0.75, how many sections must NOT be colored blue?

(A) 2

(B) 3

(C) 5

(D) 6

(E) 7

162

Ⓐ Ⓑ Ⓒ Ⓓ Ⓔ

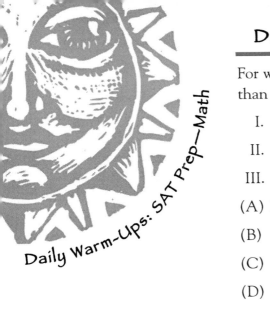

Data Analysis, Statistics, and Probability

For which of the following sets of numbers is the median greater than the mean?

   I. {2, 12, 10, 6, 8, 4}

  II. {19, 9, 1, 11, 5, 5}

 III. {15, 13, 15, 13, 15}

(A) I only

(B) II only

(C) III only

(D) I and III only

(E) II and III only

163

Data Analysis, Statistics, and Probability

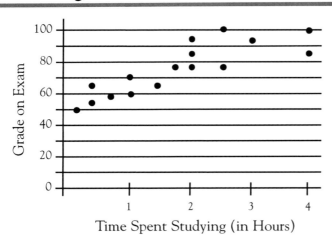

164

The scatterplot above plots the number of hours spent studying for an exam and the grades received on that exam by a group of students. What is the mode for hours spent studying?

(A) $\dfrac{1}{2}$ (B) 1 (C) 2 (D) $2\dfrac{1}{2}$ (E) 4

Ⓐ Ⓑ Ⓒ Ⓓ Ⓔ

Data Analysis, Statistics, and Probability

A certain board game involves removing a colored disk from a bag to determine how many spaces forward to move. There are red, blue, and green disks only. When the game begins, the probability of removing a red disk is $\frac{1}{2}$, the probability of removing a blue disk is $\frac{1}{6}$, and the probability of removing a green disk is $\frac{1}{3}$.

If the total number of disks is more than 25 but less than 75, what could be the total number of disks?

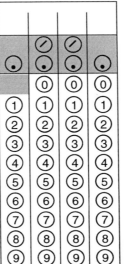

165

© 2006 Walch Publishing

Data Analysis, Statistics, and Probability

There are a total of 27 students in Mrs. Crabapple's math class. On a recent test, the girls' average score was 89, and the boys' average score was 85. If there are 15 girls in the class, what was the average score for the entire class?

(A) 88.1

(B) 87.5

(C) 87.2

(D) 86.5

(E) 86.2

166

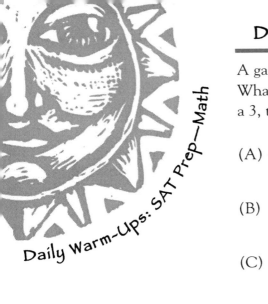

Data Analysis, Statistics, and Probability

A game player rolls a standard six-sided die six times in a row. What is the probability that the player will roll a 1, then a 2, then a 3, then a 4, then a 5 and then a 6?

(A) $\dfrac{1}{6} \times \dfrac{1}{6} \times \dfrac{1}{6} \times \dfrac{1}{6} \times \dfrac{1}{6} \times \dfrac{1}{6}$

(B) $\dfrac{1}{6} \times \dfrac{2}{6} \times \dfrac{3}{6} \times \dfrac{4}{6} \times \dfrac{5}{6} \times \dfrac{6}{6}$

(C) $\dfrac{1}{6} \times \dfrac{1}{5} \times \dfrac{1}{4} \times \dfrac{1}{3} \times \dfrac{1}{2} \times \dfrac{1}{1}$

(D) $\dfrac{1}{6} \times \dfrac{2}{5} \times \dfrac{3}{4} \times \dfrac{4}{3} \times \dfrac{5}{2} \times \dfrac{6}{1}$

(E) $\dfrac{1}{5} \times \dfrac{1}{5} \times \dfrac{1}{5} \times \dfrac{1}{5} \times \dfrac{1}{5} \times \dfrac{1}{5}$

Ⓐ Ⓑ Ⓒ Ⓓ Ⓔ

167

Data Analysis, Statistics, and Probability

The figure at the right consists of two squares. The ratio of the side length of the larger square to the smaller square is 1:3. If a point is chosen at random from the larger square, what is the probability that the point chosen will be in the small square?

(A) $\dfrac{2}{3}$

(B) $\dfrac{1}{3}$

(C) $\dfrac{2}{9}$

(D) $\dfrac{1}{8}$

(E) $\dfrac{1}{9}$

(A) (B) (C) (D) (E)

168

Data Analysis, Statistics, and Probability

{12, 4, 9, 1, 8, 4}

What is the median of the set of numbers above?

(A) 4

(B) 5

(C) 6

(D) 6.33

(E) 9

Ⓐ Ⓑ Ⓒ Ⓓ Ⓔ

Data Analysis, Statistics, and Probability

If the average of 7 different positive integers is 9, what is the greatest possible value of one of these numbers?

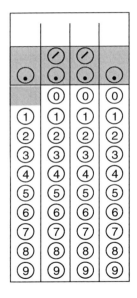

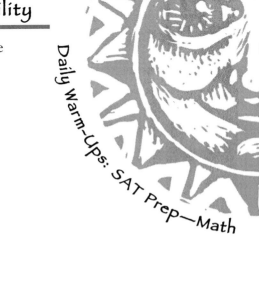

170

Data Analysis, Statistics, and Probability

A soft drink is to be chosen at random for a cooler of soft drinks.

The probability that the soft drink chosen will be ginger ale is

$\frac{3}{8}$. Which of the following could NOT be the number of soft

drinks in the cooler?

(A) 16

(B) 24

(C) 36

(D) 40

(E) 48

171

Data Analysis, Statistics, and Probability

The average of x, $\dfrac{9x}{4}$, $2x$, and a fourth number is $\dfrac{3x}{2}$. What is the fourth number?

(A) $\dfrac{3x}{4}$

(B) $\dfrac{4x}{3}$

(C) $3x$

(D) $\dfrac{17x}{3}$

(E) $\dfrac{27x}{4}$

172

Daily Warm-Ups: SAT Prep—Math

(A) (B) (C) (D) (E)

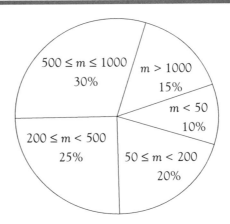

The circle graph above shows the results when 500 people in the studio audience for the taping of a program were asked, "How many miles from here do you live?" The number they gave is represented by m. How many people said that they lived less than 500 miles away?

(A) 55   (B) 150   (C) 200   (D) 225   (E) 275

Data Analysis, Statistics, and Probability

For which set of number is 13 the mode?

(A) {9, 11, 13, 15, 17}

(B) {2, 10, 20, 20}

(C) {10, 12, 14, 16}

(D) {3, 5, 11, 13, 13}

(E) {13, 13, 19, 19, 19}

174

Ⓐ Ⓑ Ⓒ Ⓓ Ⓔ

Data Analysis, Statistics, and Probability

{2, 7, 9, 11, 12, 19, 21}

What is the probability of selecting a prime number from the set of numbers above?

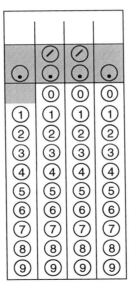

175

Data Analysis, Statistics, and Probability

If the average of a, b, and c is d, what is the value of c, in terms of a, b, and d?

(A) $a + b - d$

(B) $d - a - b$

(C) $3d - a - b$

(D) $\dfrac{a + b}{d}$

(E) $\dfrac{a + b}{3d}$

176

Daily Warm-Ups: SAT Prep—Math

   Ⓐ Ⓑ Ⓒ Ⓓ Ⓔ

Data Analysis, Statistics, and Probability

What is the average of the roots of the quadratic equation $x^2 + 5x - 6$?

(A) −2.5

(B) −0.5

(C) 0.5

(D) 2.5

(E) 3

Ⓐ　Ⓑ　Ⓒ　Ⓓ　Ⓔ

Data Analysis, Statistics, and Probability

The average of a group of consecutive multiples of 5 is 70. If the greatest of these numbers is 95, what is the smallest?

(A) 25

(B) 35

(C) 45

(D) 55

(E) 82.5

178

  Ⓐ Ⓑ Ⓒ Ⓓ Ⓔ

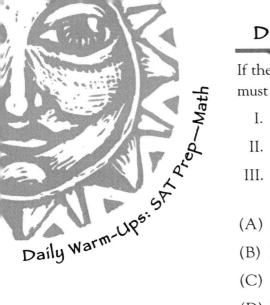

Data Analysis, Statistics, and Probability

If the median of a set of numbers is 25, which of the following must be true?

   I. The number 25 appears in the set more than any other.

  II. At least one of the numbers in the set is 25.

III. No more than half the numbers in the set are greater than 25.

(A) II only

(B) III only

(C) I and III only

(D) II and III only

(E) I, II, and III

179

    Ⓔ

Data Analysis, Statistics, and Probability

A jar can hold a maximum of 150 pieces of taffy. The probability of selecting an orange piece of taffy from the jar is $\frac{1}{5}$, and the probability of selecting a lemon piece is $\frac{1}{3}$. If there are at least 100 pieces of taffy in the jar, what could be the total number of pieces in the jar?

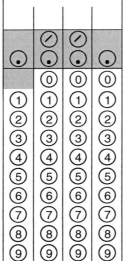

Numbers and Operations

1.	C	20.	45	39.	A
2.	B	21.	D	40.	19
3.	E	22.	D	41.	D
4.	D	23.	B	42.	B
5.	28	24.	B		
6.	D	25.	9800		
7.	B	26.	C	**Algebra and Functions**	
8.	C	27.	D	46.	D
9.	D	28.	B	47.	A
10.	13	29.	A	48.	B
11.	C	30.	90 or .9	49.	C
12.	A	31.	D	50.	9
13.	C	32.	B	51.	C
14.	B	33.	D	52.	D
15.	3, 6, 12, 24, or 48	34.	D	53.	A
16.	D	35.	1716 or 2184	54.	A
17.	A	36.	C	55.	4
18.	C	37.	D	56.	D
19.	E	38.	D	57.	E
				58.	E
				59.	D

43. D
44. C
45. 1001

60.	13		
61.	B		
62.	B		
63.	E		
64.	D		
65.	1280		
66.	C		
67.	B		
68.	C		
69.	C		
70.	1		
71.	E		
72.	A		
73.	E		

74.	B	83.	B
75.	1/2 or .5	84.	D
76.	A	85.	96
77.	B	86.	A
78.	A	87.	C
79.	D	88.	D
80.	7/2 or 3.5	89.	D
81.	A	90.	5
82.	E		

Geometry and Measurement

91.	D	101.	C
92.	C	102.	C
93.	A	103.	C
94.	D	104.	D
95.	6	105.	any number greater than 14 and less than 24
96.	C		
97.	E		
98.	B	106.	E
99.	B	107.	B
100.	10	108.	E

109.	B	123.	A
110.	13	124.	D
111.	E	125.	3
112.	D	126.	A
113.	C	127.	A
114.	D	128.	D
115.	25/8 or 3.12 or 3.13	129.	B
116.	E	130.	3
117.	C	131.	E
118.	B	132.	D
119.	C	133.	A
120.	5, 6, or 7	134.	C
121.	E	135.	81
122.	C		

Data Analysis, Statistics, and Probability

136.	B	141.	D
137.	D	142.	E
138.	E	143.	A
139.	C	144.	E
140.	36	145.	71

146.	D	163.	C
147.	E	164.	C
148.	A	165.	[30, 36, 42, 48, 54, 60, 66, or 72]
149.	B		
150.	5	166.	C
151.	E	167.	A
152.	A	168.	E
153.	C	169.	C
154.	B	170.	42
155.	1/16 or .062 or .063	171.	C
		172.	A
156.	E	173.	E
157.	C	174.	D
158.	A	175.	4/7 or .571
159.	B	176.	C
160.	1/3 or .333 or .334	177.	A
		178.	C
161.	C	179.	B
162.	A	180.	105, 120, 135, or 150

Turn downtime into learning time!

For information on other titles in the

Daily Warm-Ups series,

visit our web site: walch.com